Rain Forest Plants

Mark Numminen

REAL LIFE readers™

Rosen Classroom™

New York

Rain forests get a lot of rain.
Rain forests have many plants.

Rain forests have trees.
These trees need a lot
of rain.

Rain forests have vines.

Vines grow on trees.

Rain forests have flowers.
The flowers need a lot
of rain.

This plant grows in rain forests.

It eats bugs!

Rain Forest Plants

Words to Know

flower

rain forest

trees

vine